Crazy Chic Book

TRACK YOUR BASIC FLOCK INFO, EGG PRODUCTION, EXPENSES, COOP MAINTENANCE, HEALTH RECORDS, AND MORE WITH THIS CHICKEN KEEPING JOURNAL

3 Years

LM Taylor

MW01537591

© **Copyright 2023 - All rights reserved.**

The content contained within this book may not be reproduced, duplicated or transmitted without direct written permission from the author or the publisher.

Under no circumstances will any blame or legal responsibility be held against the publisher, or author, for any damages, reparation, or monetary loss due to the information contained within this book, either directly or indirectly.

Legal Notice:

This book is copyright protected. It is only for personal use. You cannot amend, distribute, sell, use, quote or paraphrase any part, or the content within this book, without the consent of the author or publisher.

Disclaimer Notice:

Please note the information contained within this document is for educational and entertainment purposes only. All effort has been executed to present accurate, up to date, reliable, complete information. No warranties of any kind are declared or implied. Readers acknowledge that the author is not engaged in the rendering of legal, financial, medical or professional advice. The content within this book has been derived from various sources. Please consult a licensed professional before attempting any techniques outlined in this book.

By reading this document, the reader agrees that under no circumstances is the author responsible for any losses, direct or indirect, that are incurred as a result of the use of the information contained within this document, including, but not limited to, errors, omissions, or inaccuracies.

This Book Belongs To:

_____

_____

_____

_____

Phone:_____

Email:_____

Address:_____

_____

_____

_____

Log Start Date:_____
Log End Date:_____

How to Use This Log Book

Use this chicken-keeping log to track up to three years of data related to your flock in one organized place! Track the chickens themselves, as well as egg production each month. Document different feeds given over time which can be cross-referenced to the health records as needed. And much more!

Important Contacts: Organize contact information for suppliers, breeders, veterinarians, or other people or businesses you contact regularly.

Planning Calendar: Set-up a yearly schedule for various tasks in one place for ease of reference throughout the year.

Flock Information: Use this section to track demographic information for each individual chicken. If you choose to use IDs for your birds, you can use that as a reference in other sections (egg production or health records for example).

Egg Production: Document how many eggs you get each month.

Feed Tracking: Track what kind of feed you are giving to your flock over time.

Coop Maintenance: Document when you perform different maintenance and cleaning tasks on your coop.

Health Record: Track health issues, treatments, and treatment results by bird with this section.

Income and Expenses: Track simple income and expense items related to your chicken-dom.

Summary Sheet: Transfer information to this convenient summary sheet to aggregate your data by month and year for ease of use in the future.

Notes: This section includes lined pages for you to record any other information that you wish to capture.

*****If you enjoy this chicken-keeping log book, please consider leaving a review.*****

Contacts:

Name:	
Phone:	
Address:	
Email:	
Website:	
Notes:	

Name:	
Phone:	
Address:	
Email:	
Website:	
Notes:	

Contacts:

Name:	
Phone:	
Address:	
Email:	
Website:	
Notes:	

Name:	
Phone:	
Address:	
Email:	
Website:	
Notes:	

Contacts:

Name:	
Phone:	
Address:	
Email:	
Website:	
Notes:	

Name:	
Phone:	
Address:	
Email:	
Website:	
Notes:	

Contacts:

Name:	
Phone:	
Address:	
Email:	
Website:	
Notes:	

Name:	
Phone:	
Address:	
Email:	
Website:	
Notes:	

Contacts:

Name:	
Phone:	
Address:	
Email:	
Website:	
Notes:	

Name:	
Phone:	
Address:	
Email:	
Website:	
Notes:	

Contacts:

Name:	
Phone:	
Address:	
Email:	
Website:	
Notes:	

Name:	
Phone:	
Address:	
Email:	
Website:	
Notes:	

Planning:

Month:

Month:

Month:

Planning:

Month:

Month:

Month:

Planning:

Month:

Month:

Month:

Planning:

Month:

Month:

Month:

Planning:

Month:

Month:

Month:

Planning:

Month:

Month:

Month:

Planning:

Month:

Month:

Month:

Planning:

Month:

Month:

Month:

Planning:

Month:

Month:

Month:

Planning:

Month:

Month:

Month:

Planning:

Month:

Month:

Month:

Planning:

Month:

Month:

Month:

Flock Information

ID	Name	Breed	Description	Purchase Date	DOB	Cost	Date Sold/ Death	Notes

ID	Name	Breed	Description	Purchase Date	DOB	Cost	Date Sold/ Death	Notes

ID	Name	Breed	Description	Purchase Date	DOB	Cost	Date Sold/Death	Notes

ID	Name	Breed	Description	Purchase Date	DOB	Cost	Date Sold/ Death	Notes

ID	Name	Breed	Description	Purchase Date	DOB	Cost	Date Sold/ Death	Notes

ID	Name	Breed	Description	Purchase Date	DOB	Cost	Date Sold/ Death	Notes

ID	Name	Breed	Description	Purchase Date	DOB	Cost	Date Sold/ Death	Notes

ID	Name	Breed	Description	Purchase Date	DOB	Cost	Date Sold/ Death	Notes

ID	Name	Breed	Description	Purchase Date	DOB	Cost	Date Sold/ Death	Notes

ID	Name	Breed	Description	Purchase Date	DOB	Cost	Date Sold/ Death	Notes

ID	Name	Breed	Description	Purchase Date	DOB	Cost	Date Sold/ Death	Notes

ID	Name	Breed	Description	Purchase Date	DOB	Cost	Date Sold/ Death	Notes

ID	Name	Breed	Description	Purchase Date	DOB	Cost	Date Sold/ Death	Notes

ID	Name	Breed	Description	Purchase Date	DOB	Cost	Date Sold/ Death	Notes

ID	Name	Breed	Description	Purchase Date	DOB	Cost	Date Sold/ Death	Notes

ID	Name	Breed	Description	Purchase Date	DOB	Cost	Date Sold/ Death	Notes

ID	Name	Breed	Description	Purchase Date	DOB	Cost	Date Sold/ Death	Notes

ID	Name	Breed	Description	Purchase Date	DOB	Cost	Date Sold/ Death	Notes

ID	Name	Breed	Description	Purchase Date	DOB	Cost	Date Sold/ Death	Notes

ID	Name	Breed	Description	Purchase Date	DOB	Cost	Date Sold/ Death	Notes

ID	Name	Breed	Description	Purchase Date	DOB	Cost	Date Sold/ Death	Notes

ID	Name	Breed	Description	Purchase Date	DOB	Cost	Date Sold/ Death	Notes

ID	Name	Breed	Description	Purchase Date	DOB	Cost	Date Sold/ Death	Notes

ID	Name	Breed	Description	Purchase Date	DOB	Cost	Date Sold/ Death	Notes

ID	Name	Breed	Description	Purchase Date	DOB	Cost	Date Sold/ Death	Notes

ID	Name	Breed	Description	Purchase Date	DOB	Cost	Date Sold/ Death	Notes

ID	Name	Breed	Description	Purchase Date	DOB	Cost	Date Sold/ Death	Notes

ID	Name	Breed	Description	Purchase Date	DOB	Cost	Date Sold/ Death	Notes

Records for the Month of:_____

Egg Production				Record of Feed		
Day	# of chickens	# of eggs	Notes	Week/ period	Feed	Notes
1						
2						
3						
4						
5						
6						
7						
8						
9						
10						
11				Coop Maintenance		
12				Date	Task Completed	
13						
14						
15						
16						
17						
18						
19						
20				Health Record		
21				Date	Name	Ailment
22						
23				Treatment:		
24				Date	Name	Ailment
25						
26				Treatment:		
27				Date	Name	Ailment
28						
29				Treatment:		
30				Date	Name	Ailment
31						
Tot.				Treatment:		

Notes:

Expenses			Income		
Date	Item	Cost	Date	Item	Income

Total Expense:

Total Income:

Notes:

Profit:

Records for the Month of:_____

Egg Production				Record of Feed		
Day	# of chickens	# of eggs	Notes	Week/ period	Feed	Notes
1						
2						
3						
4						
5						
6						
7						
8						
9						
10						
11				Coop Maintenance		
12				Date	Task Completed	
13						
14						
15						
16						
17						
18						
19						
20				Health Record		
21				Date	Name	Ailment
22						
23				Treatment:		
24				Date	Name	Ailment
25						
26				Treatment:		
27				Date	Name	Ailment
28						
29				Treatment:		
30				Date	Name	Ailment
31						
Tot.				Treatment:		

Notes:

Expenses			Income		
Date	Item	Cost	Date	Item	Income

Total Expense:

Total Income:

Notes:

Profit:

Records for the Month of:_____

Egg Production				Record of Feed		
Day	# of chickens	# of eggs	Notes	Week/ period	Feed	Notes
1						
2						
3						
4						
5						
6						
7						
8						
9						
10						
11				**Coop Maintenance**		
12				Date	Task Completed	
13						
14						
15						
16						
17						
18						
19						
20				**Health Record**		
21				Date	Name	Ailment
22						
23				Treatment:		
24				Date	Name	Ailment
25						
26				Treatment:		
27				Date	Name	Ailment
28						
29				Treatment:		
30				Date	Name	Ailment
31						
Tot.				Treatment:		

Notes:

Expenses			Income		
Date	Item	Cost	Date	Item	Income

Total Expense: Total Income:

Notes: Profit:

Records for the Month of:_____

	Egg Production			Record of Feed		
Day	# of chickens	# of eggs	Notes	Week/ period	Feed	Notes
1						
2						
3						
4						
5						
6						
7						
8						
9						
10						
11				Coop Maintenance		
12				Date	Task Completed	
13						
14						
15						
16						
17						
18						
19						
20				Health Record		
21				Date	Name	Ailment
22						
23				Treatment:		
24				Date	Name	Ailment
25						
26				Treatment:		
27				Date	Name	Ailment
28						
29				Treatment:		
30				Date	Name	Ailment
31						
Tot.				Treatment:		

Notes:

Expenses			Income		
Date	Item	Cost	Date	Item	Income

Total Expense:

Total Income:

Notes:

Profit:

Records for the Month of:_____

Egg Production				Record of Feed		
Day	# of chickens	# of eggs	Notes	Week/ period	Feed	Notes
1						
2						
3						
4						
5						
6						
7						
8						
9						
10						
11				Coop Maintenance		
12				Date	Task Completed	
13						
14						
15						
16						
17						
18						
19						
20				Health Record		
21				Date	Name	Ailment
22						
23				Treatment:		
24				Date	Name	Ailment
25						
26				Treatment:		
27				Date	Name	Ailment
28						
29				Treatment:		
30				Date	Name	Ailment
31						
Tot.				Treatment:		

Notes:

Expenses			Income		
Date	Item	Cost	Date	Item	Income

Total Expense: Total Income:

Notes: Profit:

Records for the Month of:_____

Egg Production				Record of Feed		
Day	# of chickens	# of eggs	Notes	Week/ period	Feed	Notes
1						
2						
3						
4						
5						
6						
7						
8						
9						
10						
11				Coop Maintenance		
12				Date	Task Completed	
13						
14						
15						
16						
17						
18						
19						
20				Health Record		
21				Date	Name	Ailment
22						
23				Treatment:		
24				Date	Name	Ailment
25						
26				Treatment:		
27				Date	Name	Ailment
28						
29				Treatment:		
30				Date	Name	Ailment
31						
Tot.				Treatment:		

Notes:

Expenses			Income		
Date	Item	Cost	Date	Item	Income

Total Expense:

Total Income:

Notes:

Profit:

Records for the Month of:_____

Egg Production				Record of Feed		
Day	# of chickens	# of eggs	Notes	Week/ period	Feed	Notes
1						
2						
3						
4						
5						
6						
7						
8						
9						
10						
11				Coop Maintenance		
12				Date	Task Completed	
13						
14						
15						
16						
17						
18						
19						
20				Health Record		
21				Date	Name	Ailment
22						
23				Treatment:		
24				Date	Name	Ailment
25						
26				Treatment:		
27				Date	Name	Ailment
28						
29				Treatment:		
30				Date	Name	Ailment
31						
Tot.				Treatment:		

Notes:

Expenses			Income		
Date	Item	Cost	Date	Item	Income

Total Expense:

Total Income:

Notes:

Profit:

Records for the Month of:_____

	Egg Production			Record of Feed		
Day	# of chickens	# of eggs	Notes	Week/ period	Feed	Notes
1						
2						
3						
4						
5						
6						
7						
8						
9						
10						
11				Coop Maintenance		
12				Date	Task Completed	
13						
14						
15						
16						
17						
18						
19						
20				Health Record		
21				Date	Name	Ailment
22						
23				Treatment:		
24				Date	Name	Ailment
25						
26				Treatment:		
27				Date	Name	Ailment
28						
29				Treatment:		
30				Date	Name	Ailment
31						
Tot.				Treatment:		

Notes:

Expenses			Income		
Date	Item	Cost	Date	Item	Income

Total Expense: **Total Income:**

Notes: **Profit:**

Records for the Month of:_____

Egg Production				Record of Feed		
Day	# of chickens	# of eggs	Notes	Week/ period	Feed	Notes
1						
2						
3						
4						
5						
6						
7						
8						
9						
10						
11				Coop Maintenance		
12				Date	Task Completed	
13						
14						
15						
16						
17						
18						
19						
20				Health Record		
21				Date	Name	Ailment
22						
23				Treatment:		
24				Date	Name	Ailment
25						
26				Treatment:		
27				Date	Name	Ailment
28						
29				Treatment:		
30				Date	Name	Ailment
31						
Tot.				Treatment:		

Notes:

Expenses			Income		
Date	Item	Cost	Date	Item	Income

Total Expense: Total Income:

Notes: Profit:

Records for the Month of:_____

Egg Production				Record of Feed		
Day	# of chickens	# of eggs	Notes	Week/ period	Feed	Notes
1						
2						
3						
4						
5						
6						
7						
8						
9						
10						
11				Coop Maintenance		
12				Date	Task Completed	
13						
14						
15						
16						
17						
18						
19						
20				Health Record		
21				Date	Name	Ailment
22						
23				Treatment:		
24				Date	Name	Ailment
25						
26				Treatment:		
27				Date	Name	Ailment
28						
29				Treatment:		
30				Date	Name	Ailment
31						
Tot.				Treatment:		

Notes:

Expenses			Income		
Date	Item	Cost	Date	Item	Income

Total Expense:

Total Income:

Notes: Profit:

Records for the Month of:_____

Egg Production				Record of Feed		
Day	# of chickens	# of eggs	Notes	Week/ period	Feed	Notes
1						
2						
3						
4						
5						
6						
7						
8						
9						
10						
11				Coop Maintenance		
12				Date	Task Completed	
13						
14						
15						
16						
17						
18						
19						
20				Health Record		
21				Date	Name	Ailment
22						
23				Treatment:		
24				Date	Name	Ailment
25						
26				Treatment:		
27				Date	Name	Ailment
28						
29				Treatment:		
30				Date	Name	Ailment
31						
Tot.				Treatment:		

Notes:

Expenses			Income		
Date	Item	Cost	Date	Item	Income

Total Expense: Total Income:

Notes: Profit:

Records for the Month of:_____

Egg Production				Record of Feed		
Day	# of chickens	# of eggs	Notes	Week/ period	Feed	Notes
1						
2						
3						
4						
5						
6						
7						
8						
9						
10						
11				Coop Maintenance		
12				Date	Task Completed	
13						
14						
15						
16						
17						
18						
19						
20				Health Record		
21				Date	Name	Ailment
22						
23				Treatment:		
24				Date	Name	Ailment
25						
26				Treatment:		
27				Date	Name	Ailment
28						
29				Treatment:		
30				Date	Name	Ailment
31						
Tot.				Treatment:		

Notes:

Expenses			Income		
Date	Item	Cost	Date	Item	Income
Total Expense:			Total Income:		
Notes:			Profit:		

Records for the Month of:_____

Egg Production				Record of Feed		
Day	# of chickens	# of eggs	Notes	Week/ period	Feed	Notes
1						
2						
3						
4						
5						
6						
7						
8						
9						
10						
11				Coop Maintenance		
12				Date	Task Completed	
13						
14						
15						
16						
17						
18						
19						
20				Health Record		
21				Date	Name	Ailment
22						
23				Treatment:		
24				Date	Name	Ailment
25						
26				Treatment:		
27				Date	Name	Ailment
28						
29				Treatment:		
30				Date	Name	Ailment
31						
Tot.				Treatment:		

Notes:

Expenses			Income		
Date	Item	Cost	Date	Item	Income

Total Expense:

Total Income:

Notes:

Profit:

Records for the Month of:_____

Egg Production				Record of Feed		
Day	# of chickens	# of eggs	Notes	Week/ period	Feed	Notes
1						
2						
3						
4						
5						
6						
7						
8						
9						
10						
11				Coop Maintenance		
12				Date	Task Completed	
13						
14						
15						
16						
17						
18						
19						
20				Health Record		
21				Date	Name	Ailment
22						
23				Treatment:		
24				Date	Name	Ailment
25						
26				Treatment:		
27				Date	Name	Ailment
28						
29				Treatment:		
30				Date	Name	Ailment
31						
Tot.				Treatment:		

Notes:

Expenses			Income		
Date	Item	Cost	Date	Item	Income

Total Expense:

Total Income:

Notes:

Profit:

Records for the Month of:_____

Egg Production				Record of Feed		
Day	# of chickens	# of eggs	Notes	Week/ period	Feed	Notes
1						
2						
3						
4						
5						
6						
7						
8						
9						
10						
11				Coop Maintenance		
12				Date	Task Completed	
13						
14						
15						
16						
17						
18						
19						
20				Health Record		
21				Date	Name	Ailment
22						
23				Treatment:		
24				Date	Name	Ailment
25						
26				Treatment:		
27				Date	Name	Ailment
28						
29				Treatment:		
30				Date	Name	Ailment
31						
Tot.				Treatment:		

Notes:

Expenses			Income		
Date	Item	Cost	Date	Item	Income
Total Expense:			Total Income:		
Notes:			Profit:		

Records for the Month of:_____

	Egg Production			Record of Feed		
Day	# of chickens	# of eggs	Notes	Week/ period	Feed	Notes
1						
2						
3						
4						
5						
6						
7						
8						
9						
10						
11				Coop Maintenance		
12				Date	Task Completed	
13						
14						
15						
16						
17						
18						
19						
20				Health Record		
21				Date	Name	Ailment
22						
23				Treatment:		
24				Date	Name	Ailment
25						
26				Treatment:		
27				Date	Name	Ailment
28						
29				Treatment:		
30				Date	Name	Ailment
31						
Tot.				Treatment:		

Notes:

Expenses			Income		
Date	Item	Cost	Date	Item	Income

Total Expense: Total Income:

Notes: Profit:

Records for the Month of:_____

	Egg Production			Record of Feed		
Day	# of chickens	# of eggs	Notes	Week/ period	Feed	Notes
1						
2						
3						
4						
5						
6						
7						
8						
9						
10						
11				Coop Maintenance		
12				Date	Task Completed	
13						
14						
15						
16						
17						
18						
19						
20				Health Record		
21				Date	Name	Ailment
22						
23				Treatment:		
24				Date	Name	Ailment
25						
26				Treatment:		
27				Date	Name	Ailment
28						
29				Treatment:		
30				Date	Name	Ailment
31						
Tot.				Treatment:		

Notes:

Expenses			Income		
Date	Item	Cost	Date	Item	Income

Total Expense: Total Income:

Notes: Profit:

Records for the Month of:_____

Egg Production				Record of Feed		
Day	# of chickens	# of eggs	Notes	Week/ period	Feed	Notes
1						
2						
3						
4						
5						
6						
7						
8						
9						
10						
11				**Coop Maintenance**		
12				Date	Task Completed	
13						
14						
15						
16						
17						
18						
19						
20				**Health Record**		
21				Date	Name	Ailment
22						
23				Treatment:		
24				Date	Name	Ailment
25						
26				Treatment:		
27				Date	Name	Ailment
28						
29				Treatment:		
30				Date	Name	Ailment
31						
Tot.				Treatment:		

Notes:

Expenses			Income		
Date	Item	Cost	Date	Item	Income

Total Expense:

Total Income:

Notes:

Profit:

Records for the Month of:_____

Egg Production				Record of Feed		
Day	# of chickens	# of eggs	Notes	Week/ period	Feed	Notes
1						
2						
3						
4						
5						
6						
7						
8						
9						
10						
11				**Coop Maintenance**		
12				Date	Task Completed	
13						
14						
15						
16						
17						
18						
19						
20				**Health Record**		
21				Date	Name	Ailment
22						
23				Treatment:		
24				Date	Name	Ailment
25						
26				Treatment:		
27				Date	Name	Ailment
28						
29				Treatment:		
30				Date	Name	Ailment
31						
Tot.				Treatment:		

Notes:

Expenses			Income		
Date	Item	Cost	Date	Item	Income

Total Expense:

Total Income:

Notes:

Profit:

Records for the Month of:_____

	Egg Production			Record of Feed		
Day	# of chickens	# of eggs	Notes	Week/ period	Feed	Notes
1						
2						
3						
4						
5						
6						
7						
8						
9						
10						
11				Coop Maintenance		
12				Date	Task Completed	
13						
14						
15						
16						
17						
18						
19						
20				Health Record		
21				Date	Name	Ailment
22						
23				Treatment:		
24				Date	Name	Ailment
25						
26				Treatment:		
27				Date	Name	Ailment
28						
29				Treatment:		
30				Date	Name	Ailment
31						
Tot.				Treatment:		

Notes:

Expenses			Income		
Date	Item	Cost	Date	Item	Income

Total Expense: | | | Total Income: | |

Notes: | Profit:

Records for the Month of:_____

	Egg Production			Record of Feed		
Day	# of chickens	# of eggs	Notes	Week/ period	Feed	Notes
1						
2						
3						
4						
5						
6						
7						
8						
9						
10						
11				Coop Maintenance		
12				Date	Task Completed	
13						
14						
15						
16						
17						
18						
19						
20				Health Record		
21				Date	Name	Ailment
22						
23				Treatment:		
24				Date	Name	Ailment
25						
26				Treatment:		
27				Date	Name	Ailment
28						
29				Treatment:		
30				Date	Name	Ailment
31						
Tot.				Treatment:		

Notes:

Expenses			Income		
Date	Item	Cost	Date	Item	Income

Total Expense:

Total Income:

Notes:

Profit:

Records for the Month of:_____

Egg Production				Record of Feed		
Day	# of chickens	# of eggs	Notes	Week/ period	Feed	Notes
1						
2						
3						
4						
5						
6						
7						
8						
9						
10						
11				Coop Maintenance		
12				Date	Task Completed	
13						
14						
15						
16						
17						
18						
19						
20				Health Record		
21				Date	Name	Ailment
22						
23				Treatment:		
24				Date	Name	Ailment
25						
26				Treatment:		
27				Date	Name	Ailment
28						
29				Treatment:		
30				Date	Name	Ailment
31						
Tot.				Treatment:		

Notes:

Expenses			Income		
Date	Item	Cost	Date	Item	Income

Total Expense:

Total Income:

Notes:

Profit:

Records for the Month of:_____

Egg Production				Record of Feed		
Day	# of chickens	# of eggs	Notes	Week/ period	Feed	Notes
1						
2						
3						
4						
5						
6						
7						
8						
9						
10						
11				**Coop Maintenance**		
12				Date	Task Completed	
13						
14						
15						
16						
17						
18						
19						
20				**Health Record**		
21				Date	Name	Ailment
22						
23				Treatment:		
24				Date	Name	Ailment
25						
26				Treatment:		
27				Date	Name	Ailment
28						
29				Treatment:		
30				Date	Name	Ailment
31						
Tot.				Treatment:		

Notes:

Expenses			Income		
Date	Item	Cost	Date	Item	Income
Total Expense:			Total Income:		
Notes:			Profit:		

Records for the Month of:_____

Egg Production				Record of Feed		
Day	# of chickens	# of eggs	Notes	Week/ period	Feed	Notes
1						
2						
3						
4						
5						
6						
7						
8						
9						
10						
11				Coop Maintenance		
12				Date	Task Completed	
13						
14						
15						
16						
17						
18						
19						
20				Health Record		
21				Date	Name	Ailment
22						
23				Treatment:		
24				Date	Name	Ailment
25						
26				Treatment:		
27				Date	Name	Ailment
28						
29				Treatment:		
30				Date	Name	Ailment
31						
Tot.				Treatment:		

Notes:

Expenses			Income		
Date	Item	Cost	Date	Item	Income

Total Expense:

Total Income:

Notes:

Profit:

Records for the Month of:_____

Egg Production				Record of Feed		
Day	# of chickens	# of eggs	Notes	Week/ period	Feed	Notes
1						
2						
3						
4						
5						
6						
7						
8						
9						
10						
11				**Coop Maintenance**		
12				Date	Task Completed	
13						
14						
15						
16						
17						
18						
19						
20				**Health Record**		
21				Date	Name	Ailment
22						
23				Treatment:		
24				Date	Name	Ailment
25						
26				Treatment:		
27				Date	Name	Ailment
28						
29				Treatment:		
30				Date	Name	Ailment
31						
Tot.				Treatment:		

Notes:

Expenses			Income		
Date	Item	Cost	Date	Item	Income
Total Expense:			Total Income:		
Notes:			Profit:		

Records for the Month of:_____

Egg Production				Record of Feed		
Day	# of chickens	# of eggs	Notes	Week/ period	Feed	Notes
1						
2						
3						
4						
5						
6						
7						
8						
9						
10						
11				Coop Maintenance		
12				Date	Task Completed	
13						
14						
15						
16						
17						
18						
19						
20				Health Record		
21				Date	Name	Ailment
22						
23				Treatment:		
24				Date	Name	Ailment
25						
26				Treatment:		
27				Date	Name	Ailment
28						
29				Treatment:		
30				Date	Name	Ailment
31						
Tot.				Treatment:		

Notes:

Expenses			Income		
Date	Item	Cost	Date	Item	Income

Total Expense: **Total Income:**

Notes: **Profit:**

Records for the Month of:_____

	Egg Production			Record of Feed		
Day	# of chickens	# of eggs	Notes	Week/ period	Feed	Notes
1						
2						
3						
4						
5						
6						
7						
8						
9						
10						
11				Coop Maintenance		
12				Date	Task Completed	
13						
14						
15						
16						
17						
18						
19						
20				Health Record		
21				Date	Name	Ailment
22						
23				Treatment:		
24				Date	Name	Ailment
25						
26				Treatment:		
27				Date	Name	Ailment
28						
29				Treatment:		
30				Date	Name	Ailment
31						
Tot.				Treatment:		

Notes:

Expenses			Income		
Date	Item	Cost	Date	Item	Income

Total Expense:

Total Income:

Notes:

Profit:

Records for the Month of:_____

Egg Production				Record of Feed		
Day	# of chickens	# of eggs	Notes	Week/ period	Feed	Notes
1						
2						
3						
4						
5						
6						
7						
8						
9						
10						
11				Coop Maintenance		
12				Date	Task Completed	
13						
14						
15						
16						
17						
18						
19						
20				Health Record		
21				Date	Name	Ailment
22						
23				Treatment:		
24				Date	Name	Ailment
25						
26				Treatment:		
27				Date	Name	Ailment
28						
29				Treatment:		
30				Date	Name	Ailment
31						
Tot.				Treatment:		

Notes:

Expenses			Income		
Date	Item	Cost	Date	Item	Income

Total Expense: Total Income:

Notes: Profit:

Records for the Month of:_____

	Egg Production			Record of Feed		
Day	# of chickens	# of eggs	Notes	Week/ period	Feed	Notes
1						
2						
3						
4						
5						
6						
7						
8						
9						
10						
11				Coop Maintenance		
12				Date	Task Completed	
13						
14						
15						
16						
17						
18						
19						
20				Health Record		
21				Date	Name	Ailment
22						
23				Treatment:		
24				Date	Name	Ailment
25						
26				Treatment:		
27				Date	Name	Ailment
28						
29				Treatment:		
30				Date	Name	Ailment
31						
Tot.				Treatment:		

Notes:

Expenses			Income		
Date	Item	Cost	Date	Item	Income
Total Expense:			Total Income:		
Notes:				Profit:	

Records for the Month of:_____

Egg Production				Record of Feed		
Day	# of chickens	# of eggs	Notes	Week/ period	Feed	Notes
1						
2						
3						
4						
5						
6						
7						
8						
9						
10						
11				**Coop Maintenance**		
12				Date	Task Completed	
13						
14						
15						
16						
17						
18						
19						
20				**Health Record**		
21				Date	Name	Ailment
22						
23				Treatment:		
24				Date	Name	Ailment
25						
26				Treatment:		
27				Date	Name	Ailment
28						
29				Treatment:		
30				Date	Name	Ailment
31						
Tot.				Treatment:		

Notes:

Expenses			Income		
Date	Item	Cost	Date	Item	Income

Total Expense: | | Total Income:

Notes: | Profit:

Records for the Month of:_____

Egg Production				Record of Feed		
Day	# of chickens	# of eggs	Notes	Week/period	Feed	Notes
1						
2						
3						
4						
5						
6						
7						
8						
9						
10						

				Coop Maintenance		
11						
12				Date	Task Completed	
13						
14						
15						
16						
17						
18						
19						

				Health Record		
20						
21				Date	Name	Ailment
22						
23				Treatment:		
24				Date	Name	Ailment
25						
26				Treatment:		
27				Date	Name	Ailment
28						
29				Treatment:		
30				Date	Name	Ailment
31						
Tot.				Treatment:		

Notes:

Expenses			Income		
Date	Item	Cost	Date	Item	Income

Total Expense:

Total Income:

Notes:

Profit:

Records for the Month of:_____

	Egg Production			Record of Feed		
Day	# of chickens	# of eggs	Notes	Week/ period	Feed	Notes
1						
2						
3						
4						
5						
6						
7						
8						
9						
10						
11				Coop Maintenance		
12				Date	Task Completed	
13						
14						
15						
16						
17						
18						
19						
20				Health Record		
21				Date	Name	Ailment
22						
23				Treatment:		
24				Date	Name	Ailment
25						
26				Treatment:		
27				Date	Name	Ailment
28						
29				Treatment:		
30				Date	Name	Ailment
31						
Tot.				Treatment:		

Notes:

Expenses			Income		
Date	Item	Cost	Date	Item	Income

Total Expense:		Total Income:	
Notes:		Profit:	

Records for the Month of:_____

Egg Production				Record of Feed		
Day	# of chickens	# of eggs	Notes	Week/ period	Feed	Notes
1						
2						
3						
4						
5						
6						
7						
8						
9						
10						
11				**Coop Maintenance**		
12				Date	Task Completed	
13						
14						
15						
16						
17						
18						
19						
20				**Health Record**		
21				Date	Name	Ailment
22						
23				Treatment:		
24				Date	Name	Ailment
25						
26				Treatment:		
27				Date	Name	Ailment
28						
29				Treatment:		
30				Date	Name	Ailment
31						
Tot.				Treatment:		

Notes:

Expenses			Income		
Date	Item	Cost	Date	Item	Income
Total Expense:			Total Income:		
Notes:				Profit:	

Records for the Month of:_____

Egg Production				Record of Feed		
Day	# of chickens	# of eggs	Notes	Week/ period	Feed	Notes
1						
2						
3						
4						
5						
6						
7						
8						
9						
10						
11				Coop Maintenance		
12				Date	Task Completed	
13						
14						
15						
16						
17						
18						
19						
20				Health Record		
21				Date	Name	Ailment
22						
23				Treatment:		
24				Date	Name	Ailment
25						
26				Treatment:		
27				Date	Name	Ailment
28						
29				Treatment:		
30				Date	Name	Ailment
31						
Tot.				Treatment:		

Notes:

Expenses			Income		
Date	Item	Cost	Date	Item	Income

Total Expense:

Total Income:

Notes:

Profit:

Records for the Month of:_____

Egg Production				Record of Feed		
Day	# of chickens	# of eggs	Notes	Week/ period	Feed	Notes
1						
2						
3						
4						
5						
6						
7						
8						
9						
10						
11				**Coop Maintenance**		
12				Date	Task Completed	
13						
14						
15						
16						
17						
18						
19						
20				**Health Record**		
21				Date	Name	Ailment
22						
23				Treatment:		
24				Date	Name	Ailment
25						
26				Treatment:		
27				Date	Name	Ailment
28						
29				Treatment:		
30				Date	Name	Ailment
31						
Tot.				Treatment:		

Notes:

Expenses			Income		
Date	Item	Cost	Date	Item	Income

Total Expense: Total Income:

Notes: Profit:

Records for the Month of:_____

Egg Production				Record of Feed		
Day	# of chickens	# of eggs	Notes	Week/ period	Feed	Notes
1						
2						
3						
4						
5						
6						
7						
8						
9						
10						
11				Coop Maintenance		
12				Date	Task Completed	
13						
14						
15						
16						
17						
18						
19						
20				Health Record		
21				Date	Name	Ailment
22						
23				Treatment:		
24				Date	Name	Ailment
25						
26				Treatment:		
27				Date	Name	Ailment
28						
29				Treatment:		
30				Date	Name	Ailment
31						
Tot.				Treatment:		

Notes:

Expenses			Income		
Date	Item	Cost	Date	Item	Income
Total Expense:			Total Income:		
Notes:			Profit:		

Summary Data, Year:_____

Month	# of chickens	# of eggs	# Hatched	# Deaths	# Bought	# Sold	Income	Expense	Profit
Jan									
Feb									
Mar									
Apr									
May									
Jun									
Jul									
Aug									
Sep									
Oct									
Nov									
Dec									
Total									

Notes:

Summary Data, Year:_____

Month	# of chickens	# of eggs	# Hatched	# Deaths	# Bought	# Sold	Income	Expense	Profit
Jan									
Feb									
Mar									
Apr									
May									
Jun									
Jul									
Aug									
Sep									
Oct									
Nov									
Dec									
Total									

Notes:

Summary Data, Year:_____

Month	# of chickens	# of eggs	# Hatched	# Deaths	# Bought	# Sold	Income	Expense	Profit
Jan									
Feb									
Mar									
Apr									
May									
Jun									
Jul									
Aug									
Sep									
Oct									
Nov									
Dec									
Total									

Notes:

Notes

Notes

Notes

Notes

Notes

Notes

Notes

Notes

Notes

Notes

Notes

Notes

Notes

Notes

Notes

Notes

Notes

Notes

Notes

Notes

Made in the USA
Coppell, TX
26 November 2024

41065441R10090